Guilherme Mapa
Matheus Nascimento

Socio-environmental Analysis with an Emphasis on Water Resources

Guilherme Mapa
Matheus Nascimento

Socio-environmental Analysis with an Emphasis on Water Resources

Study on the upper and middle reaches of the Córrego dos Macacos in Belo Horizonte-MG

ScienciaScripts

Cover image: www.ingimage.com

This book is a translation from the original published under ISBN 978-613-9-68726-8.

Publisher:
Sciencia Scripts
is a trademark of
Dodo Books Indian Ocean Ltd. and OmniScriptum S.R.L publishing group

120 High Road, East Finchley, London, N2 9ED, United Kingdom
Str. Armeneasca 28/1, office 1, Chisinau MD-2012, Republic of Moldova, Europe
Printed at: see last page
ISBN: 978-620-8-22920-7

SUMMARY

WHEN I WALKED ALONG THE TRAILS OF GRANJA WERNECK, NOW APPROACHING THE SOLIMOES NEIGHBORHOOD, WE IMAGINED, AS INNOCENT CHILDREN, WALKING THROUGH A FOREST FULL OF TUCANOS, TIÙS, MICOS AND BEAUTIFUL FLORA.

The process of urbanization evolved over more than 20 years and some of the fauna sought refuge wherever there was a minimum of conditions for survival. We wanted to do something to preserve their natural habitat, but to no avail. In a meeting with Geography, we understood the relationship between society and nature, and in this work we saw an opportunity to seek a balance in this relationship.

In the face of many obstacles, we managed to realize our dream and we thank God for giving us the health and strength to overcome the difficulties.

To this university, its teaching staff, management and administration, who provided the window through which we can now glimpse a higher horizon.

To our advisor Professor Dr. João Henrique Rettore and the Cartography laboratory technician, Professor Ms. Gustavo Libério de Paula, for their support in the short time they had, for their corrections and encouragement.

To our parents, for their love, encouragement and unconditional support.

INTRODUCTION

The capital of Minas Gerais, Belo Horizonte, which was conceived and planned along modernist lines, was born with segregation projects in its urban space. The city's planning provided for the delimitation of areas for different social classes, with the Avenida do Contorno delimiting the boundary of the urban area. Throughout the 1960s, 1970s and 1980s, the city's growth and the development of industrialization began to attract rural populations in search of better living conditions. Faced with this reality, this homeless population began to occupy the urban voids irregularly in search of a place to live, further accentuating the process of socio-spatial segregation. In this context of poverty and inequality, socially segregated segments emerge, without access to basic urban infrastructure services and with limited access to social services such as health, education and leisure (CAIADO, 1998 *apud* GARCIA & RIBEIRO, 2001).

The process of socio-spatial segregation continues in the city, especially in vectors of expansion such as the northern region of the capital. This region has attracted investment from the state government and the municipality of Belo Horizonte, as well as private enterprise, increasing the value of land. The metropolitan center of Venda Nova, which articulates an extensive area with high population growth, constitutes a space in which there is a process of reproduction of peripheries, housing those excluded from the process of metropolization. With the increase in local real estate speculation and the government's neglect of the vector's growth process, social movements began to take shape, which, in search of housing, ended up occupying empty areas of the axis, such as Granja Werneck.

Located in the northern region of Belo Horizonte, Granja Werneck (also known as Mata do Isidoro) is an area of approximately 950 hectares that is home to a large green area native to the city. Its history begins with the granting of the area by the city council to the doctor Hugo Werneck in 1914 for the construction of a sanatorium to treat tuberculosis in the region, which was inaugurated in 1928. From the 1970s onwards, the sanatorium became an asylum (Recanto da Boa Viagem) and was taken over by the Catholic Church through FONSBEM - Fundaçâo Obras Sociais Nossa Senhora da Boa Viagem. The local landscape also includes an old quarry set up in the 1950s and various neighborhoods established since the 1970s, which border the area, both from Belo Horizonte and Santa Luzia. The area is currently home to three major urban occupations in the struggle for the right to housing, and is a source of conflict between the region's occupying residents, the state and private enterprise.

The area of occupation is around 3.6 million square meters and there is a plan for intervening urban action influenced by the expansion works of the northern vector of Belo Horizonte. The Granja Werneck Project, according to the advertising and planning released by Direcional Engenharia, aims to install a sustainable urbanization project there.

The academic sources available for gathering information on the history of the occupations are still few, as the occupations are relatively new. For the most part, there are reports and documentaries usually published by social organizations supporting the occupations, demonstrations on social networks, blogs and official notes from the city council.

The communities - Rosa Leao, Vitória and Esperança - have an official blog on which their slogans read: "The occupations in the Isidora region, in Belo Horizonte and Santa Luzia, MG, were formed during the June 2013 protests. Today there are around 8,500 landless and homeless families who can no longer tolerate the heavy burden of rent. They are fighting for the right to decent housing. Follow our struggle through our blog and, if possible, come and visit us and commit yourself to our struggle, which is a just, legitimate and constitutional one."

In 2014, an official note was issued by the Belo Horizonte City Hall requesting the repossession of the area, but this was not carried out, but at all times the families live in fear of this action. In addition to their irregular nature, the occupations are accused of damaging green areas and water resources, causing environmental stress in the area, which has extensive cerrado and Atlantic Forest vegetation.

In view of this brief physical, historical and social context, what are the problems faced by the local population with regard to water impacts on the surface and groundwater levels in the Córrego dos Macacos watershed? How does the community deal with the environment? What means are needed to solve the problems arising from the relationship between man and nature?

In view of the above questions, this work seeks to carry out an analysis of the selected spatial area in order to identify and qualify the relationship between man and nature, especially with regard to water resources in the Córrego dos Macacos watershed. It also aims to point out some solutions to reduce the environmental impact in the area. The aim is also to provide a physical characterization of the study area; to understand the production of the geographic space selected; to understand how environmental impacts occur and how anthropic action can intensify this process; to point out some solutions, minimizing the environmental impact in the area; and to diagnose the perception of local residents in relation to the environment and their degree of involvement with it.

The importance of environmental preservation for the maintenance and conservation of soil, water resources and springs is fundamental for the perpetuation of human life. When there is any kind of interference with the environment, mitigating solutions must be sought that can significantly contribute to a balance between society and nature.

As such, the region under study, which has a privileged location, is on the route of urban and real estate expansion, but is considered one of Belo Horizonte's remaining green areas and is directly threatened by the agents that produce urban space. As such, this study is important for surveying the impacts that have occurred in the micro-basin of the Macacos stream, as well as the importance of the area as a water recharge zone, since it is home to several springs and watercourses that feed important river basins, such as the Isidoro and Onça streams, in Belo Horizonte and the metropolitan region.

Saraiva (2016) cites that the main interventions for the main problems in Granja Werneck require the specific activities of a large professional body and that geographers are responsible for diagnoses, questionnaire application and raising awareness among the population, recognizing and studying physical and geographic characteristics, mapping, geoprocessing and cartography, assessing the potential of water and environmental resources, delimiting and characterizing regions for systemic planning.

In this way, this work will help the local population to have arguments to strengthen their demands for regularization and sustainable urbanization, as well as supporting other studies and projects aimed at minimizing the environmental impact *in situ* and in the surrounding areas.

Although there are three major settlements in the area that makes up Granja Werneck, the Esperança settlement, located in the north, was delimited because it includes the sources of the Córrego dos Macacos - a contributor to the Ribeirâo Isidoro and, consequently, to the Ribeirâo do Onça and Rio das Velhas. Analyzing the quality of the water at this point is essential for contributing to public health - not only in the Esperança occupation, but also throughout the Isidoro basin, as well as having other implications for monitoring the quality of the soil throughout the region.

1. THEORETICAL BACKGROUND

1.1 The production of the Belo Horizonte metropolis

With the advent of the Republic, Ouro Preto, the capital of Minas Gerais since 1720, no longer had the support to be the political and development center of the state. Thus, in a new capital, urban life and values tended to favor republican practice.

Belo Horizonte emerged with the aim of being a modern city inspired by European urbanism, obtaining suitable conditions due to its topography, climate, hydrographic basin and geographical location. It was built in 1893 and has delimited areas with more modernized urban areas for the elite, suburban areas and rural areas.

Figure 1. Cadastral Plan of the City of Minas - 1895

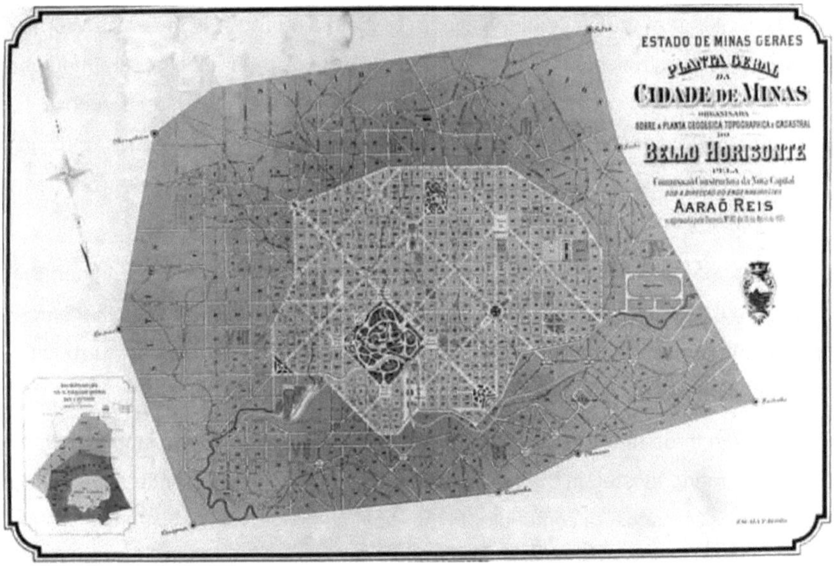

Source: Public Archive of the City of Belo Horizonte

The area within the limits of the Avenida do Contorno was considered urban and thus destined for public facilities and housing for the elite. This avenue served as a frontier, separating urban life from suburban life, where housing was inferior and precarious. It also separated the green belt that supplied the city with food.

The process of Belo Horizonte's metropolization was brought about by a developmentalist and inflationary economic policy, with an increase in industrialization. However, the absence of coherent social policies resulted in a rapid process of demographic growth, establishing a perverse and accelerated situation of slumification.

According to Machado Silva (2002):

> "[...] the favela won. But what about the favela dwellers? [...] [they remain] bearers of a restricted, hierarchical and fragmented citizenship [...] the 'victory of the favela' occurred at the cost of the constitution of a subaltern category, whose hard-won intervention on the public scene did not change the pattern of urban society, hardly altering its relative position in social stratification and its role as a social force." (MACHADO SILVA, 2002)

The rigidity of the city plan, which earmarked the area inside the Avenida do Contorno perimeter for specific functions, led the working classes to occupy the suburban areas of the city. As a result, this population occupied slopes and hilltops with a risky geomorphological profile. In contrast to the city's design, it lost its hygienic character, lacking basic sanitation, adequate roads and urban equipment.

Belo Horizonte, as a city created for modernity, presupposed urban planning as a priority and, consequently, legislation to order and control the appropriation of urban land. What we saw, therefore, was a reverse process.

In the 1930s, there was an urban crisis resulting from the accelerated growth of the city as a result of a speculative real estate market. The process of urban occupation was characterized primarily by the suburban space, which had already been identified since the capital's beginnings. The phenomenon was even stimulated by the public administration itself, which imposed strict legislation on the appropriation of the central area, restrictions not found with such rigor in the suburban area.

In this way, the urban center was surrounded by an accelerated occupation characterized by a tangle of villages, forming an extensive, sparse urban fabric whose infrastructure services did not meet the demands of the population.

The new urban order was established through the economic order, without due concern for the social order, since the expropriated lands became commodities whose monetary value had no real estate correspondence in the new city. This unforeseen issue, of incongruous real estate values, led to the expulsion of the inhabitants of the arraial, leading them to occupy the outskirts of the new city, heralding the miserable suburbs that would characterize Brazilian cities in the last century.

1.2 The urban space

According to Santos (1997), all spaces are geographical because they are determined by the movement of society, of production. But both landscape and space are the result of society's movements, in other words, they are the fruit of human actions on space itself, through natural and/or artificial means.

Space can have various functions that are defined according to the use assigned to it. Urban space tends to concentrate a series of services to offer from the center to the periphery. Urban space is the materialization of modern society.

According to Corrêa (1993), urban space is a set of different land uses juxtaposed to each other. The various uses of this land constitute the spatial organization of the city. Urban space is both fragmented and articulated, since it maintains spatial relations with other areas, varying its intensity significantly.

In this way, urban space is produced by a series of social factors. According to Corrêa (1993), the agents who produce urban space are: the owners of the means of production (mainly industrialists), landowners, real estate developers, the state and excluded social groups. The actions of these agents serve the dominant purpose of capitalist society, which is the reproduction of production relations, which implies the continuity of the accumulation process and the attempt to minimize class conflicts, a role generally assumed by the state. In this way, the relations of production in this space take place between the flow of people, cars, and the transportation of goods, among others.

Based on the relationships that produce space, we have territoriality which, according to Araùjo (2012), is a system of place and ways of identifying and differentiating oneself within it, outlining one's own way of life and preserving habits, customs and ways of living together and expressing oneself.

In this way, urban border spaces are places of contrast characterized by their appropriation by different social groups. According to Souza (2010), as a result, it is common for uses to be different, or even alternate over time. In this way, public space creates an inexhaustible complexity in the context of life experiences and produces an interaction between people with different purposes who transform these spaces according to their uses.

In his work "From Confederates to Barbarians", Maldi (1997) tries to reconstruct an approach to territoriality and the border as cultural categories in which he looks for a "basic resonance between the object and the way it is perceived, since they have to be understood, in the first place, from the representation that society makes of itself and the other, that is, how it defines territoriality and how and with whom it defines its borders."

Given that urban space is a reflection of society, it reveals segregation, not only of housing, but also of services, commerce and even leisure areas. We can clearly see the unequal class structure to which the urban environment is subjected and how urban infrastructure is closely linked to this spatial organization.

According to Corrêa (1993):

> Fragmented, articulated, reflecting and conditioning society, the city is also the place where the various social classes live and reproduce themselves. This involves everyday life and the near future, as well as the beliefs, values and myths created within class society and, in part, projected onto spatial forms: monuments, sacred places, a special street, etc. Urban space thus takes on a symbolic dimension which, however, varies according to the different social and age groups, etc. But everyday life and the near future are framed in a context of unequal fragmentation of space, leading to social conflicts (...). The space of the city is thus also the setting and object of social struggles, since these are ultimately aimed at the right to the city, full and equal citizenship for all. (CORRÊA, 1993. p. 9)

According to the author, the producers of urban space act by putting pressure on the state to favor their interests. For example, landowners can intervene in municipal laws that regulate land use, seeking to increase the value of their property and thus make more profit. In general terms, it can be said that in a capitalist society, the social agents who produce urban areas are always looking for the right conditions to expand their capital or profits.

In this way, the urban landscape is indicative of its history and conditions of development, revealing the weight of the past in the organization of urban space in contemporary times. By observing urban evolution, one can follow the changes in social content, so that the city reveals in its various spatial elements the socio-economic contrasts of its inhabitants. (DOLLFUS, 1975)

It can therefore be said that each place/space has unique specificities and that it is these that generate in individuals a sense of belonging, of origin.

According to Santos (1997), place is:

> [...] a set of objects that have autonomy of existence due to the things that make them up - streets, buildings, pipes, industries, businesses, restaurants, electricity, sidewalks, but which do not have autonomy of meaning, because every day new functions replace the old ones, new functions are imposed and exercised. (SANTOS, 1997. P. 52)

In this way, place is confused with space, but according to Tuan (2013) "Space is more abstract than place". What begins as undifferentiated space becomes place as we get to know it better and give it value.

Along the same lines, Geiger (1998) quotes Lefebvre:

> A concept is produced by immediate sensations, psychologically, and by logic, logically. Containing limits of sovereignty, property, discipline, surveillance and jurisdiction, the territory psychologically conveys the sensation of enclosure (Lefebvre, 1947 *apud* Geiger, 1998).

According to Santos (1997), space is the result of human action on space itself, mediated by natural and artificial objects. Spatialization is a moment in the territorial insertion of social processes that can be expressed in the form of a psychological barrier - which we call psycho-territorial here - and can then be perceived in the landscape. With all its form and spatial organization, social formation is reflected in the landscape. According to the author, forms are both the result and the condition of the process. Spatial forms are not passive, but active.

We seek the possibility of achieving basic urban infrastructure, information, technology, coexistence, leisure, social and political participation with fair income distribution, benefits guaranteed by the constitution and described in the City Statute of 2001, one of the alternatives which is the Specific Global Plan (PGE), which consists of:

"[...] a planning instrument that guides urban restructuring, environmental and social development interventions in towns, slums and low income housing estates. It consists of an in-depth study of the reality of these areas, taking into account urban planning, socio-economic aspects and the legal situation of the land. The main objective of the PGE is to point out ways of improving the quality of life in these areas and integrating them into the city as a whole (PBH, PGE URBEL 2015)".

Recognition and urban and legal intervention actions are the first fundamental steps.

Integrated planning is an instrument of Belo Horizonte's city hall that aims to serve villages and slums within the municipality. For some, it is considered a modern project to decentralize urban investment towards villages and slums, the Municipal Housing Policy. These programs use the municipality's own resources, as well as external resources from the federal government and international organizations, in order to make the goals set by the popular movements more effective. In order to implement the PGE (Specific Global Plan), it is necessary to collect data and make an environmental, legal, economic and organizational diagnosis in order to make the best proposals for social, legal and urban environmental interventions through community mobilization. (URBEL PBH 2015)

Among the Municipal Housing Policies with similar demands to the community under study, the structural intervention of the Vila Viva program stands out as a reference. This program has already benefited several communities and slums in the capital, such as Conjunto da Serra, Taquaril, Vila Acaba Mundo and others. This intervention has a structuring character that promotes profound transformations in the existing settlement, with the aim of integrating it into the city and providing adequate housing conditions.

1.3 . Watershed and micro-watershed

The study of the hydrological cycle and river basins is made up of interconnections and tends to consider various factors in its dynamics. Although there are conceptualizations in this regard, emphasis should be placed on human relations, the physical constitution of the relief and the climate, among others, so as not to dissociate one element from another in order to understand the interactions involved in this issue.

The concept of the river basin as a unit of analysis and management must be linked to the concept of sustainable development with a view to emphasizing social equity, economic and environmental development, thus fulfilling the function of the river basin from a geographical perspective.

Apart from the various approaches under which the watershed can be studied, according to the different areas, what must be considered is the spatial scale; in this regard, Nacif (1997) *apud* Schiavetti and Camargo (2002) points out that watersheds are units that can be considered true "cells" whose sum gives rise to the "fabric" called the earth's surface. The components of these "cells" are natural resources and humans, who, through society, represent the real agents in these units.

Borannn and Likens (1967) *apud* Schiavetti and Camargo (2002) consider the watershed to be a basic ecosystem unit. Considering the various phases of the water cycle, this is where the greatest degree of anthropogenic interference occurs. However, the use of water as a "resource" for human activities generally disregards the fact that we are dealing with a system in which the parts interrelate according to a complex and delicate dynamic. The latter author emphasizes the urgency of using a systems approach in the management of freshwater and urbanized areas, where this perception is hampered by the high degree of alteration of the landscape.

Various works allow us to understand a hydrographic basin as a geographically defined region on the territory as a cut-out of the surface. It encompasses the drainage network including a main river and its tributaries, causing all the water precipitated there to flow into a single outlet - a cross-section of the main river - called the outflow or control section of the watershed, according to Silva (2015, p. 57).

For the National Policy for Water Resources Management in Brazil (Law 9433/1997), a hydrographic basin is understood as a territorial unit for the implementation of the national water resources policy and the action of the national water resources management system, thus being a political and administrative issue.

According to Silva (2015), the hydrological cycle is a global phenomenon of closed circulation of water between the earth's surface, the subsoil and the atmosphere, driven by solar energy, associated with gravity and the earth's rotation. The cycle has no strictly defined beginning or end and the various processes involved take place continuously and dynamically; the main processes that make it up are therefore precipitation, runoff, storage (surface and underground), evapotranspiration and condensation.

For Tucci (1997), the hydrographic basin is made up of a set of surfaces, slopes and a drainage network formed by watercourses that converge until they result in a single bed at their outlet. Horton (1945) made a significant contribution to the definition of the basin as an individualized but hierarchical system when he identified parameters for the organization of drainage channels.

From another perspective, watersheds can also be understood as:

> "[...] open systems, which receive energy through climatic agents and lose energy through defluxion, can be described in terms of independent variables, which oscillate around a pattern and thus, even when disturbed by anthropic actions, are in dynamic equilibrium." (LIMA & ZAKIA, 2000).

In view of the various concepts that seek to define the term *watershed*, based on different and complementary parameters, we understand that it can be approached as a fragment of geographical space delimited by higher areas that perform the functions of creating and recharging aquifers, as well as surface water drainage. This fragment is characterized by specific climatic, pedological, geological and biological conditions, which are dynamically related and can suffer from anthropic intervention over time.

When it comes to the concept of a micro-basin, according to Faustino (2009), we can understand that it is an area that drains directly into the main course of a sub-basin. In this way, we can consider that the Córrego dos Macacos is a micro-basin of the Ribeirao Isidoro, with an area of less than 100 km .^{2}

From the point of view of hydrology, the classification of watersheds into large and small is not only seen in terms of their total surface area, but also taking into account the effects of certain dominant factors on the generation of runoff. Therefore, this explanation contributes to the distinction, definition and spatial delimitation of micro-basins and watersheds, and its understanding is crucial for structuring environmental monitoring, by means of field measurements for water sampling, topography and cartography and with the help of Geographic Information Systems - GIS (TEODORO; et al. 2007).

1.4 . Environmental vulnerability and water resources

Vulnerability involves a set of factors that can decrease or increase the risk(s) to which human beings, individually or as a group, are exposed in the various situations of their lives. These situations can be, for example, a flood, a landslide, an expense, social marginalization, an economic recession, (ESTEVES, 2011).

Environmental vulnerability or fragility is seen in an integrated way and is related to an area's susceptibility to damage when subjected to a particular action. According to Coelho (2010), mapping tends to portray spatialization, i.e. a process of formation and differentiated spatial, temporal and social distribution of environmental impacts.

According to Jansen (2013), any integrated analysis of a natural environment presupposes an understanding of the dynamics of its functioning, with human intervention. Environmental systems are more or less fragile depending on their 'in-form' characteristics, i.e. some kind of alteration to the different components of the nature of physical geography (relief, soil, vegetation, climate and water resources) that could compromise the functioning of an entire system, breaking its state of dynamic equilibrium.

Natural processes such as erosion, sediment deposition in bodies of water, mass movements, karst sinkholes, separation of pollutants in groundwater and aquifer recharge have their intensity associated with factors such as climate, soil types (associated with rock types), slope and vegetation cover and may or may not be subject to human interference (MENDONÇA, 2004).

Thus, it can be understood that the area under study naturally has an environmental vulnerability due to its physical characteristics - such as soil, topography and plant material - which intensifies with anthropogenic action. According to Garcias and Sanches (2009), with regard to impacts on water sources, these have direct consequences both for the natural environment, by altering the natural characteristics of the source, and for human society, by altering the quality of the water and making it difficult to use it for supply, and can have other indirect consequences.

The main issues linked to water in the urban environment are related to water pollution, lack of sewage treatment, occupation along the riverbed, leading to frequent flooding, waterproofing and canalization of urban rivers, whether for motorways or due to lack of public awareness.

According to Tucci (2008, p. 101) in developing countries, cities vary in stage. Initially, when the population is small, the supply is made from wells or a nearby body of water, and the sewage is discharged into the drainage system (when it exists) or goes into black wells or septic tanks, with the possibility of the water supply being contaminated by the sewage itself.

This scenario is dramatic when the soil has low infiltration capacity. The cesspools don't work and the sewage flows into the streets or down drains. This stage predates hygiene, which causes the proliferation of water-borne diseases.

In this pre-hygienist stage, diseases such as diarrhea are the main cause of infant mortality and Brazil is not yet at the hygienist stage due to the lack of sewage treatment, the transfer of flooding in drainage and the lack of solid waste control.

In order to understand the reality of the springs and watercourses present in the micro-basin of the Córrego dos Macacos, which is within the perimeter of Granja Werneck, it is necessary to consult the literature on current legislation in relation to water treatment, water quality and urban hydrology.

According to Von Sperling (2005),

"Man's interference, whether in concentrated form, such as the generation of domestic or industrial waste, contributes to the introduction of compounds into water, affecting its quality. Therefore, the way in which man uses and occupies the land has a direct impact on water quality." (SPERLING, 2005, p.16)

Water pollution is a worrying factor for humanity, and can be defined as any damage to its properties that directly and adversely affects man as a living organism, or indirectly by reducing the value of its physical objects or properties, or even opportunities for recreation or appreciation of nature.

According to the National Environmental Policy (PNMA, 1981) pollution is:

"All environmental degradation resulting from activities that directly or indirectly harm the health, safety and well-being of the population, create adverse conditions for social and economic activities, adversely affect the biota, affect the aesthetic or sanitary conditions of the environment, release materials or energy in disagreement with established environmental standards, environmental degradation, alteration of the characteristics of the environment". (PNMA, 1981)

The quality of water is linked to its use, so it is essential to study its quality in order to establish certain uses and analyze pollution factors. The quality of water is based on various aspects since, as a solvent and its transport capacity, it incorporates residues from natural and man-made conditions. In addition to the hydrological cycle, it also takes into account the factors of external cycles whose characteristics are altered as a result of its use in various sectors such as residential, reuse, rainwater, industrial and others. The impurities in water can be solid or organic and its analysis is carried out as chemical, biological and physical.

According to Sperling (2006, p. 22), the classification of solids present in water is attributed to their physical characteristics, which are: suspended solids, colloidal solids and dissolved solids. These parameters can be defined by color, turbidity, temperature, taste and odor. As for the chemical characteristics, they are classified as organic and inorganic, being: hydrogenionic potential (pH), alkalinity, acidity, hardness, iron, manganese, chlorides, nitrogen, phosphorus, dissolved oxygen, organic matter, inorganic micropollutants and organic micropollutants (see Annex A). For water quality analysis, biological parameters include organisms such as viruses, bacteria and protozoa, which are part of the coliform group normally found in animal feces, water and soil.

"Escherichia coli (E. coli): the main bacterium of the fecal coliform group, abundant in human and animal feces: guarantee that contamination is exclusively fecal; easy detection in the laboratory." (CETESB, 2015.)

When it comes to types of pollution, Sperling (2006, p. 48) cites two types: point source pollution, when the pollutants reach the body of water in a concentrated way in space, citing as an example the discharge of a river from an outfall carrying sewage from a community; and diffuse pollution, which refers to pollutants that enter the body of water, being distributed along part of its length (a typical example is pollution that has rainwater drainage as its vehicle, being distributed and not concentrated at a single point).

In order to determine the use of water, its classification is analyzed. The CONAMA resolution (1986, no. 20) classifies the use of water into stages according to its function. Water distributed as special class is intended for domestic supply without prior or simple disinfection and the preservation of the natural balance of aquatic communities.

Table 1: Classification criteria - CONAMA Resolution (1986)

CLASSIFICATION OF WATER USE

HYDRIC CLASS	CHARACTERISTICS
Class 1	Intended for domestic supply after simplified treatment, protection of aquatic communities, primary contact recreation, irrigation of vegetables and fruit grown close to the ground and consumed raw, and agriculture of species intended for human consumption.
Class 2	Intended for domestic supply after conventional treatment for the protection of aquatic communities, primary contact recreation, irrigation of plants and vegetables and fruit plants and in agriculture.
Class 3	Less demanding for certain uses, they are intended for domestic supply after conventional treatment, irrigation of tree, cereal and fodder crops and animal watering.

Created by the authors, based on CONAMA (1986)

1.5 Environmental risks and impacts

Environmental impacts can be negative or positive. They can occur without man even having set foot in a given space, in other words, they suffer the consequences of interference caused by distant "epicenters".

According to CONAMA (1986), environmental impacts are:

> [...] any alteration of the physical, chemical and biological properties of the environment, caused by any form of matter or energy resulting from human activities which, directly or indirectly, affects: the health, safety and well-being of the population; social and economic activities; biota; the aesthetic and sanitary conditions of the environment; the quality of environmental resources. (BRASIL, 1986)

Moreira (1997) defines environmental impact as a powerful influence exerted on the environment, causing the natural ecosystem to become unbalanced. What characterizes environmental impact is not any change in the properties of the environment, but changes that cause an imbalance in the constitutive relationships of the environment, such as changes that exceed the absorption capacity of the environment in question.

This opposition is revealed as: (a) any alteration to the properties of the environment and (b) explains that it is not any alteration to the environment and the level of damage must be greater than the environment's capacity to regenerate. However, it should be borne in mind that there are natural impacts and that, over time, nature filters them out or builds another landscape, but in the short or medium term they generate impacts for human beings, such as natural fires that increase the low humidity of the air, the discharge of suspended particulates, hurricanes, among other natural phenomena.

With the process of urbanization and industrialization, the physical environment has historically suffered anthropic intervention, whether in the "countryside" or in cities. The "natural" environment and the anthropologically constructed environment reveal the transformations that have taken place in the landscape in favor of economic development, generating environmental impacts that reflect on the quality of life of the population and biodiversity. Environmental impact is related to a cause/interference that generates positive or negative consequences, since anthropic activity has a decisive influence on the biosphere and consequently on the habitat itself. Humans use nature as a means of obtaining the raw materials needed to produce consumer goods for the benefit and reproduction of the capitalist system. Risk is driven by the representation of the imminent possibility, or not, that something could happen.

The impact brings the risk, for example, of compromised food production due to the impact on the soil through impoverishment or desertification as a result of management practices, and can thus be associated with various processes of damage to the environment, often linked to each other. There is, of course, some likelihood of other risks, such as the extinction of springs in a hydrographic basin due to the impact of plant suppression, erosion, siltation and others to meet the needs of formal and informal housing without environmental or socio-environmental support.

According to Egler (1996) *apud* Castro, Peixoto & Rio (2007), natural risk is associated with the behavior of natural systems, considering the degree of stability and instability expressed by vulnerability to events of short or long duration. According to White et al. (2001), these are intrinsically linked to the use of natural resources and the transformation of sites by society.

For Foucher (1982), natural risks increase with population growth and, on a local scale, they increase with the urbanization of often vulnerable sites (floodplains, low-lying regions, foothills, etc.). Faced with the environmental impacts and risks that human activity produces, it leads to unsustainable living conditions in the short and long term.

1.6 Land use and occupation

Given that urban space is a reflection of society, it reveals segregation, not only of housing, but also of services, commerce and even leisure areas. We can clearly see the unequal class structure to which the urban environment is subjected and how urban infrastructure is closely linked to this spatial organization.

According to Corrêa (1993):

> Fragmented, articulated, reflecting and conditioning society, the city is also the place where the various social classes live and reproduce themselves. This involves everyday life and the near future, as well as the beliefs, values and myths created within class society and, in part, projected onto spatial forms: monuments, sacred places, a special street, etc. Urban space thus takes on a symbolic dimension which, however, varies according to the different social groups, ages, etc. But everyday life and the near future are framed in a context of unequal fragmentation of space, leading to social conflicts (...). The space of the city is thus also the scene and object of social struggles, since these are ultimately aimed at the right to the city, full and equal citizenship for all. (CORRÊA, 1939. p. 9)

Land use and occupation should be seen as urban planning mechanisms. It can be said that the term "land use and occupation" is defined in terms of the rules relating to densification, the regime of activities, building control devices and land parceling, which make up the urban planning regime (TAKEDA, 2013).

The purpose of the term is to control population density and land occupation by buildings, to improve commuting and urban mobility, to eliminate the possibility of environmental disasters and to preserve the environment. To achieve this goal, tools are needed to enable and/or guarantee urban planning that focuses on a mix of interests. Among these tools are the Master Plan and Zoning.

The Master Plan is the basic instrument of urban policy for municipalities, with the aim of promoting appropriate land use planning, as well as the full development of the city's social functions and guaranteeing the well-being of its inhabitants. For such planning, it is necessary to control the use, subdivision and occupation of land in compliance with the guidelines of the Statute of Cities. Its purpose is to guide the actions of the public authorities and the private sector in the construction of urban, rural and industrial spaces and in the provision of essential public services, with the aim of ensuring better living conditions for the population. In addition, the Master Plan must provide for the delimitation of urban areas where compulsory subdivision, building or use may be applied, taking into account the existence of infrastructure and demand for use.

Zoning is another tool of fundamental importance in city planning, guaranteeing its orderly development. In it, the municipal territory is divided into parts (called zones) where land use and occupation rules are defined for each one. The territory of the Municipality of Belo Horizonte is considered to be an urban area and is divided into Zones according to their characteristics or potential (art. 4 of Law 7.166/96).

The municipality's zoning is delimited in Annex II of Law 7.166/96 (art. 14 of Law 7.166/96). The Zones or Zoning are differentiated according to the possibilities of building and population density and the demands of environmental, historical, cultural, archaeological or landscape preservation and protection. For each zoning district, criteria are established for land subdivision, implantation and use of buildings (figure 2) called "urban planning parameters". (art. 4 of Law 7.166/96)

Figure 2 - Zoning of Granja Werneck - 1996

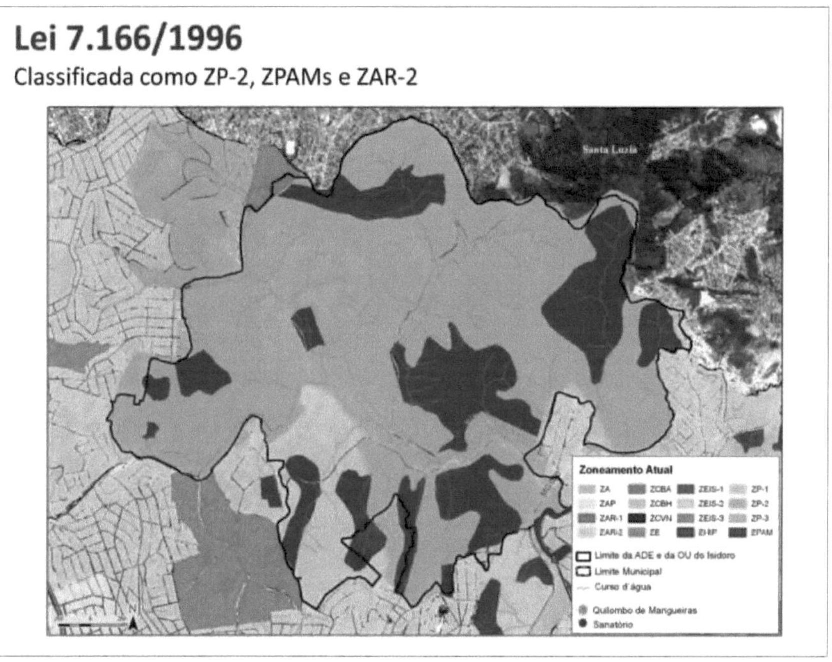

Source: PBH (2010)

The specificities and particularities of some regions are taken into account in the Areas of Special Guidelines - ADEs, Special Areas of Social Interest - AEIS - and in proposals for Urban Operations, when additional and/or differentiated criteria for parceling, occupation and specific use are established.

An amendment to the LPOUS (Land Development, Occupation and Use Law) in 2010 (Law 9959/2010) made changes to the zoning map (figure 3) of Granja Werneck in order to comply with the urban planning project defined for the area, as well as defining the ADE of the Isidoro region as an integrated set of interventions with the aim, according to the legislation, to promote the protection and environmental recovery of the Isidoro Region, through a process of orderly and sustainable occupation, which provides for the preservation of areas of great environmental and landscape importance, especially springs and watercourses, and areas of significant vegetation and savannah.

Figure 3 - Zoning (Northern part of Granja Werneck)

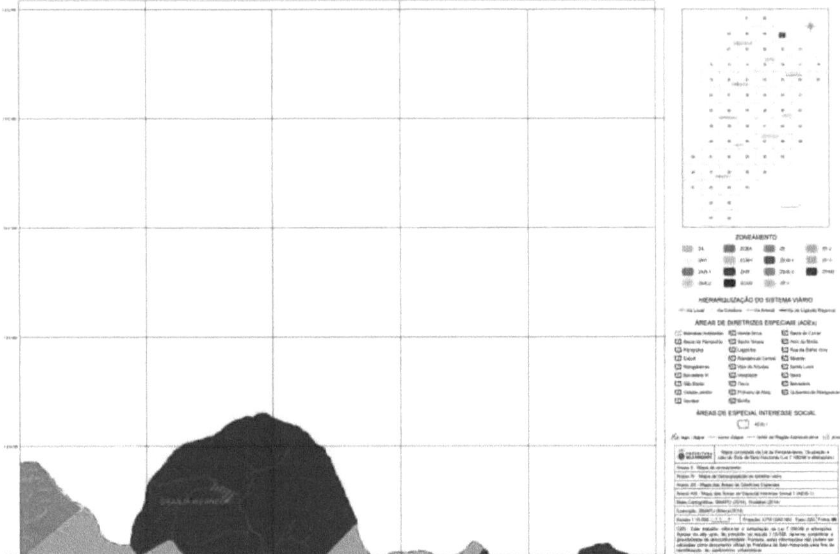

Source: PBH (2010)

The private sector owns large areas of land in the region and acts by putting pressure on governments to change the municipal laws that regulate land use, seeking to increase the value of their property and thus make more profit. In general terms, it can be said that in a capitalist society the main social agents who produce urban areas are always looking for the right conditions to expand their capital or their profits. The state manifests itself at various points, firstly through the PBH (Belo Horizonte City Hall), which changed the zoning of the capital, taking into account the construction of the large development in the region.

The ADE do Isidoro, previously delimited by the PBH at 3.6 million square meters, was, according to a report published by Revista Ecológica, extended to the whole of Granja Werneck, adding up to 9.5 million square meters, an area larger than that of old Belo Horizonte, 8.5 million, dreamed up by Aarao Reis, within the perimeter of Avenida do Contorno. This is due to the northern growth vector, which has increased in value as a result of the installation of important facilities such as the Estaçao BH shopping mall and the Presidente Tancredo Neves Administrative City, among others.

Despite the legal specifications and government interventions on the issue of land use and occupation in Belo Horizonte, in practice, according to Santos (2008), land use and occupation already have recipients. The best subdivisions are built in the planned areas with the best infrastructure, while popular subdivisions are built on the fringes of the city, strengthening the process of spatial segregation of the population in the municipality and the RMBH. According to the author:

In its contemporary version, technology has been put at the service of production on a planetary scale, where neither the limits of states, resources nor human rights are taken into account. Nothing is taken into account except the unbridled pursuit of profit, wherever the elements capable of enabling it can be found (SANTOS, 2008).

When the 2010 law was amended, it defined three urban planning categories and their parameters, with areas that have relevant vegetation, topography and water resources that are prone to environmental recovery being reclassified by CONAMA (National Environment Council). These three categories were defined as follows:

Table 2: Categories - CONAMA Resolution (1986)

URBAN CATEGORIES (CONAMA)

GRADE	CHARACTERISTICS
Degree of Protection 1	Areas of maximum protection, intended for the permanent preservation of springs, watercourses and large continuous areas of vegetation cover of environmental importance, where occupation should be prohibited, except for activities related to their maintenance and preservation.
Degree of Protection 2	Areas of high protection due to topographical conditions, the significant presence of watercourses and isolated patches of significant vegetation cover, in which occupation, densification and soil sealing must be restricted.
Degree of Protection 3	Moderate protection areas, in which, due to the more favorable local topographical, morphological and drainage conditions and the lower concentration of relevant vegetation cover, the following parameters may be established occupation and densification are less restricted than in other areas.

Created by the authors, based on CONAMA (1986)

In the capital of Minas Gerais, throughout its construction there has been an intense process of land use and occupation to cater for a public with a higher socio-economic level, marked by the launch of a large number of housing developments in the Belo Horizonte Metropolitan Region (RMBH), mainly in Nova Lima.

> Land speculation is therefore not in the interests of the owners of the means of production. It is, however, in the interest of landowners: the retention of land creates a shortage of supply and an increase in its price, enabling them to increase their income from the land (CORRÊA, 1993, p. 14).

The quality of allotments became the main strategy of the real estate market to control the use and occupation of land in the capital and the RMBH, because through its direct influence on the price of the plot, the real estate market selected the inhabitants of the more upmarket regions of the capital, expelling the population with a lower socio-economic level to the outskirts of the municipality itself or to other nearby municipalities. This process explains the conurbation between Belo Horizonte and Santa Luzia.

The issue of land use and occupation in Granja Werneck, despite having a project aimed at its orderly and segregated occupation, ended up being the scene of a social struggle for housing. The intervention of the Minas Gerais State Public Prosecutor's Office led to delays in the licensing process, which prevented the project from starting. During the course of the process, the social movements took action, occupying the large area, restricting the state release, since there are people living in the area. There are three large occupations in Granja Werneck, including the Esperança occupation, a community that is close to the source of the Córrego dos Macacos, an area zoned as ZEPAM by the Belo Horizonte City Hall.

Private interests and the inefficiency of public authorities in regularizing the occupation and incorporating its residents into the social fabric end up neglecting the social right to decent housing. Integrating the occupations through the parameters defined by the LPOUS can collaborate with the sustainable development of the residents, guaranteeing the right to housing and acting as a promoter of an orderly and conscientious occupation process.

2. METHODOLOGY

In order to achieve the research objectives, a theoretical foundation was laid for the previously selected concepts that will contribute to this work. The theoretical foundation was the product of reading and analyzing fragments of books, academic articles, master's dissertations and doctoral theses. The concepts selected were: watershed, vulnerability, risk and environmental impact, as well as a review of legislation on land use and occupation, and the parameters indicated for water analysis.

After the theoretical foundation, a brief physical characterization of the study area was carried out, based on bibliographic studies on climatic and structural geomorphology, analysis of satellite images from the last ten years to compare degraded areas, as well as the use of GIS (Geographic Information Systems), especially ArcGIS *software*, to create cartographic materials. Maps were drawn up showing the location, hydrography and water sampling points.

To create the thematic maps, we first searched the databases of the Brazilian Institute of Geography and Statistics (IBGE), where it is possible to *download* topographic maps. We selected sheet SE-23-Z-C-VI-3, georeferenced and available on a scale of 1:50,000, which covers the territory of the municipality of Belo Horizonte (MG). Subsequently, the quadrant covering the Granja Werneck area was located and vectorized to serve as the basis for all the maps.

In addition to vectorizing the study area using the topography map, the geographic location/position map was drawn up using *shapefires* (year 2013) made available by PRODABEL - Belo Horizonte City Hall. The hydrology map will be made on the basis of the location map, also using the base *(shpefire)* of the Minas Gerais Institute of Water Management (IGAM*)*. The study area was empirically defined using a Google Earth image and then transferred to ArcGIS *software.*

The historical recovery of human occupation in the region will take place by collecting information through the accounts of the oldest residents, recording videos and texts, and this data will be analyzed considering the visual alteration of the landscape. This change will take place through the visualization of satellite images made available on Google Earth (timeline tool) and images generated from *drone* flights, made available by the Office of Architecture and Urbanism at PUC Minas.

In addition to characterizing the study area, an assessment of the current water quality situation will be carried out at previously strategically defined points, using some of the parameters employed by the Minas Gerais State Water Management Institute (IGAM). Georeferenced samples were taken at points upstream of the main urbanization strip.

Figure 4: Collection points

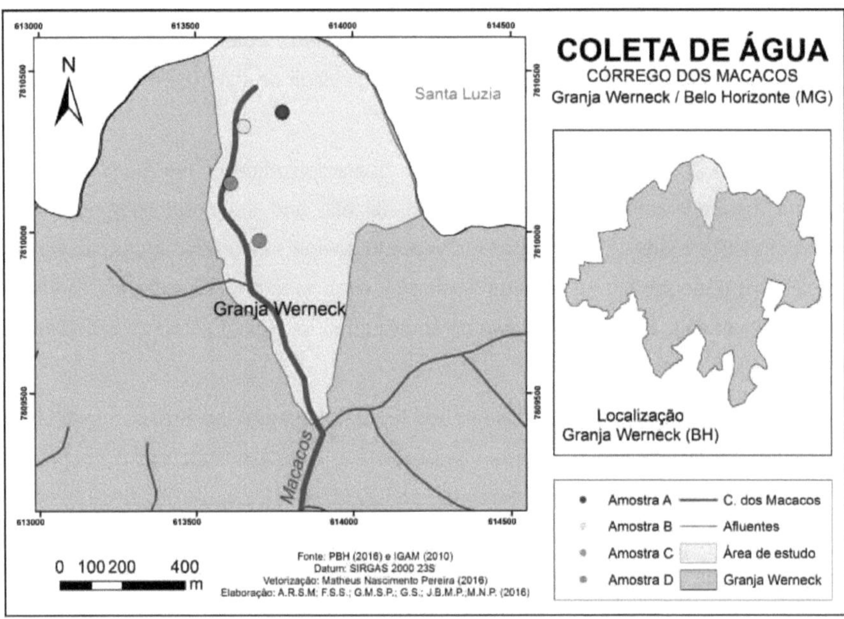

The samples were collected in May and November 2017, using 500 ml polyethyleneflalate containers previously sterilized with HNO3. PH, temperature, TDS and conductivity were analyzed using Analytical Instruments 986 and AKSO AK87 8602 multi-parameter readers. These parameters were selected on the basis of the possibilities currently available to the group for analyzing water samples.

The samples will be stored and transported in a Styrofoam box until they are properly analyzed in the laboratories of the Department of Geography and/or the Postgraduate Program in Geography - Treatment of Spatial Information at PUC Minas.

In order to carry out and understand the perception of the residents of the area under study, this study will use the interview technique which, according to Rosa and Arnoldi (2006) and Luna (1988) *apud* Jûnior A. and Jûnior N. (2011), "refers to a data collection technique considered to be a rational form of conduct by the researcher, previously established, to efficiently direct a systematic content of knowledge [...]." For Ribeiro (2008) *apud* Jûnior A. and Jûnior N. (2011), the interview is a pertinent technique when the researcher wants to obtain information about the object, which allows them to learn about attitudes, feelings and values underlying behavior, going beyond descriptions of actions and incorporating new sources so that the results can be properly interpreted.

We therefore opted to use semi-structured interviews, which were quick and relatively inexpensive, and we believe that this type of interview will be more open, allowing the interviewee to speak more freely, but without the group losing focus. The questions will be prepared in advance and we will use handwritten forms to conduct the interview, as well as recorders available on *smartphones*. The recordings will be listened to, transcribed using text editing *software* (Microsoft Office Word) and incorporated with other information, enabling a more comprehensive analysis of the perception of the social groups living near the Monkeys stream, in relation to the impacts caused by anthropic actions and the outlook for the future in relation to social inclusion.

In a small sample of people, we collected information such as age, previous place of residence, current job, number of residents per household, length of time living in the community, the origin of the water used and its disposal, priorities and prospects for the future.

3. CHARACTERIZATION OF THE REGION

3.1 Location and geographical position

Granja Werneck is located between latitudes 19° 49' 5.45" S and 19° 48' 44.31" S; and longitudes 43° 54' 56.24" W and 43° 54' 51.95" W. It is located in the northern region of the municipality of Belo Horizonte, in Minas Gerais, and its territory borders the metropolitan region that includes the municipality of Santa Luzia.

Figure 5 Location map of the study area

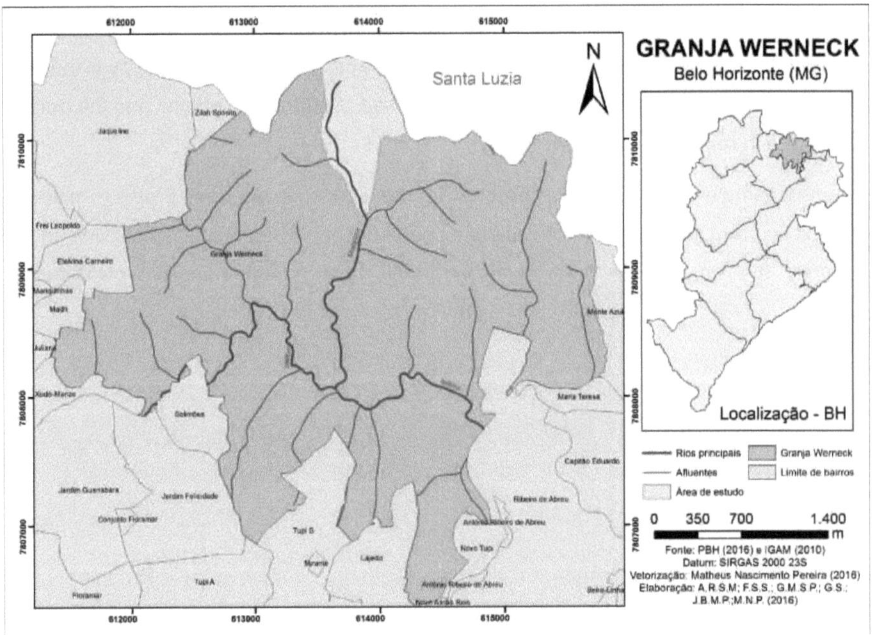

3.2 Geology and soils

According to the Geological Program of Brazil, in the Southeast region, within the state of Minas Gerais and forming part of the tectonic area of the São Francisco Craton in the region of the Iron Quadrangle, the Granja Werneck study area is part of the Minas supergroup.

The Belo Horizonte complex is made up of partially remobilized, migmatized and millonitized Paleoproterozoic Archean gneissic rocks (EIA RIMA GRANJA WERNECK, 2001, P.291). It is important to highlight the characteristics of the geology and the types of soil in the Izidoro region so that actions to interfere in this environment are carried out in a sustainable manner.

The study area is represented only by the high-grade formations of the Belo Horizonte complex, of gneiss rocks covered locally by residual soil or alluvial and colluvial sediments (EIA RIMA GRANJA WERNECK, 2001, P.291).

During the fieldwork, the analysis of aerial images from Google Earth in its timeline and the study of the EIA RIMA GRANJA WERNECK of 2001, it was possible to identify a series of anthropogenic events over the years that accelerated the processes of erosion, gullying and silting up of water courses and springs, often associated with deforestation and occupation of the headwaters of the area.

On the surface, immature, siliceous, aluminous, micaceous, acidic residual soils are often found in the millonitized zones and dark red soils are associated with basic and ultrabasic, meso to melanocratic rocks. Intrusive rocks of at least three types are present, dykes of unmetamorphized dolerite, characteristically producing clayey soils from yellow ochre to blood red, metamorphized ultramafic intrusives and quartz veins, some thick, forming globular masses visible next to the Sanatorio quarry, divided with Santa Luzia and at other points on the banks of the Izidoro (EIA RIMA GRANJA WERNECK, 2001, p.191).

The Precambrian rocks can be grouped into two lithostructural and morphological groups; the region where the Belo Horizonte complex is formed, according to the Geological Survey of Brazil Program, occupies 70% of the municipality and the others are metasedimentary (GEOLOGICAL SURVEY OF BRAZIL PROGRAMME, 2000 p.7).

The relief of the Isidoro basin can be considered moderate to strongly undulating, with spikes controlled in a north-south direction or close to it. This control, defined by the structure of the Belo Horizonte complex, also varies in the main course of the Isidoro and its tributaries (EIA RIMA GRANJA WERNECK, 2001, P.302).

In view of the previous studies carried out, we identified the study area as an area of strong environmental vulnerability due to the genetic processes of soil and landscape formation accelerated by anthropogenic interventions in the region.

3.3 Hydrology

Hydrography is a field of research that should be considered as the basis for the interrelationship between nature and society, with climate, relief and pedological structure as the pillars of support for the balance of communities and the environment.

The Rio das Velhas basin was of great historical importance for the development of the central region of Minas Gerais, having been one of the main routes through which the gold cycle developed, since it runs through several cities with these economic activities, cities that became industrialized in the region such as Sabarà, Santa Luzia and Belo Horizonte, occupying a drainage area of 27. 85 km2, between latitudes 17 15 S and 20° 25' S and longitudes 43° 25' W and 50' W. The drainage area of this basin represents 12% of the region's area.857 km2, between latitudes 17^{0} 15' S and 20° 25' S and longitudes 43° 25' W and 44° 50' W. The drainage area of this basin represents 12% of the Sâo Francisco basin in Minas Gerais. At 801 km, the Rio das Velhas is the largest tributary of the Sâo Francisco Basin. It rises in the municipality of Ouro Preto, within the recently created Andorinhas Municipal Park, and flows into the Velho Chico in the district of Barra do Guaicuy, municipality of Vàrzea da Palma.

The drainage pattern of most of the basin's watercourses is of the dendritic type, common to regions of crystalline rocks or basement rocks. The tributaries of the Rio das Velhas include, on the right bank, the Curimatai river (Buenópolis municipality), the Paraùna river, the main tributary, the Cipó river (a tributary of the Paraùna river located between the municipalities of Santana de Pirapama, Presidente Juscelino and Gouveia) and the Jaboticatubas river (Jaboticatubas municipality). On the left bank are the Cotovelo stream (municipality of Pirapora), the Bicudo stream (Corinto), the Picâo stream (Curvelo), the Onça stream (Cordisburgo) and the rivers and streams that drain the Belo Horizonte Metropolitan Region (Arrudas and Onça). The density of the natural drainage network shows greater hydrographic richness among the right bank tributaries, a fact associated with the basin's geological characteristics.

There is academic research that suggests that the Velhas River was one of the main routes for the discovery of precious metals in Minas Gerais. It is believed that explorers coming from the north-east would have gone up along the banks of the São Francisco river (which collects the waters of the Velhas basin) and then up along the Velhas river. It's a much longer route than the one from Salvador-BA, but due to the lack of roads at the time, the rivers were used as references since they also had resources for human needs, such as fishing, as well as being a more natural route. Arriving in the central region of Minas Gerais, the explorers discovered gold and precious stones, which caused wars due to disputes over the mines in the region.

The River das Velhas has suffered a series of interferences over the years. A large part of its water volume is captured for treatment and use by the population, such as the Bela Fama Water Treatment Plant; and due to the establishment of cities on its banks, the river receives a large amount of sewage through tributaries that are used for sanitary transportation, such as the Arrudas stream and the Onça stream, which cross the city of Belo Horizonte.

The metropolitan region of Belo Horizonte, which according to the master plan of the Velhas

Hydrographic Basin Committee, is in the portion called Alto Rio das Velhas and occupies only 10% of the territorial area of the basin, but has more than 70% of its population. In addition, it is where industrial activities are concentrated and where there is an advanced urbanization process, sometimes without projects to mitigate the impacts on water bodies, as is the case with occupations, making it the area that contributes most to the degradation of the waters of the Rio das Velhas.

Environmental degradation means that the Rio das Velhas has reddish-colored waters in its stretch, where there is heavy iron ore mining, due to the presence of the metal which is leached out by precipitation in areas where the vegetation layers of the region's soil have been removed, leaving the river extremely polluted and silted up. There is practically no life in the river's waters along this stretch with high population density and many industrial activities.

In 1997, a group of professors from the Federal University of Minas Gerais (UFMG) started a project to clean up the River das Velhas. With the aim of bringing the Velhas river basin back to life, the project began a series of actions to raise public awareness, which has brought results through achievements in many municipal and state public policies, and particularly with stricter controls for the pollution emitters installed along the basin. The project had the ambitious goal of revitalizing the River das Velhas by 2010.

In 2009, the Rio das Velhas Hydrographic Basin Committee carried out the 'Expediçâo pelo rio das Velhas', a project called the Manuelzâo project, which toured the entire basin holding mobilization and environmental education events.

Law No. 9.433/97 states in its first article that:

I - water is a public good; II - water is a limited natural resource with economic value; IV - the management of water resources must be decentralized and include the participation of the public authorities, users and communities.

Article 2 of this law states:

I - ensuring the necessary availability of water for current and future generations, at quality standards appropriate to their respective uses; II - the rational and integrated use of water resources, including waterway transportation, with a view to sustainable development.

Based on the guidelines, it can be seen that both parties are not meeting the requirements, which requires greater awareness on the part of the riverside dwellers and the entire community, which even though it is remote from the water resource, has an impact on it.

Not far from this reality is the ribeirâo do Onça, which is a watercourse that runs through the cities of Belo Horizonte and Contagem and flows into the river Velhas, forming a sub-basin just like the ribeirâo Arrudas. The ribeirâo do Onça is located in the Alto Velhas region, on the left bank of the river Velhas, in the most urbanized part of the basin. The Onça stream flows into the River das Velhas in the municipality of Santa Luzia.

The process of urbanization in the capital of Minas Gerais has "hidden" the region's watercourses. The discharge of sewage, the reduction of drainage areas and disorderly occupation are some of the main problems caused by the intense occupation of the sub-basins, which have damaged the conditions of the Velhas River. Trying to maintain the natural conditions of the watercourses is one of the main guidelines of Drenurbs (Environmental Recovery and Sanitation Program for Valley Bottoms and Natural Streams in Belo Horizonte). Drenurbs works to recover the city's streams that are still in their natural beds. With the installation of the Arrudas Sewage Treatment Plant - ETE Arrudas - and the ETE Onça, it is hoped that the waters of the Velhas will improve. The Arrudas plant was the first in Belo Horizonte; it opened in October 2001 and is located in the municipality of Sabarâ. The Ribeirâo da Onça Sewage Treatment Plant (ETE Onça), located in the Ribeiro de Abreu district, was inaugurated in 2006 and has the capacity to treat 1,800 liters of sewage per second, which corresponds to 155 million liters daily. The Nossa Senhora da Piedade stream, which flows directly into the Onça, has clean water and several springs, and does not receive water from any other stream. Copasa inaugurated the Onça ETE in 2006.

The Nossa Senhora da Piedade Park was created, a sports area that integrates society with nature, as well as Belo Horizonte's first Environmental Education Center. The first waters that the stream receives come from the city of Contagem and form the Pampulha lagoon, then it passes along Avenida Cristiano Machado and receives the waters of several streams: Gorduras, Ouro Minas, Guarani, Aarâo Reis, Saramenha, and finally the waters of the Isidoro stream, which receives waters from the Venda Nova region, Serra Verde and Cidade Administrativa.

The Isidoro basin, where the Córrego dos Macacos is located, is under an area of influence since it has the limits of basins and sub-basins that function in an integrated way with the rocky substrate, soils and watercourses.

Figure 6 Hydrography of the region

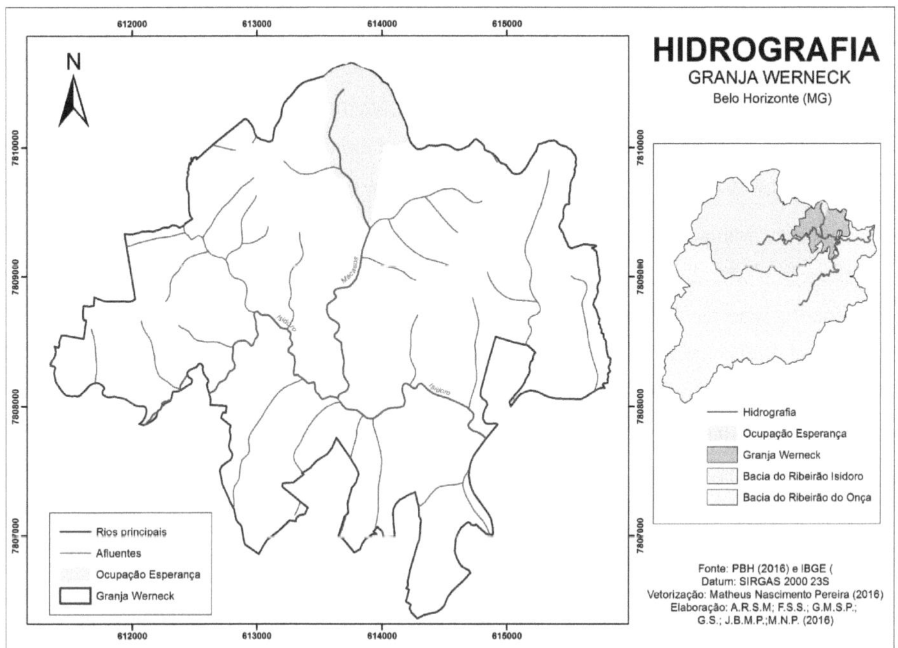

The Isidoro stream, which is in one of the few empty areas of Belo Horizonte and also one of the last with much vegetation cover. The capital is completely built up, with only 4% of unoccupied areas left, a fact that impacts on the flow of precipitated water and percolation into the water table. Granja Werneck and other areas of Ribeirao Isidoro account for 90% of these remaining areas. The Córrego dos Macacos is considered the last clean watercourse in Belo Horizonte.

Throughout the Isidoro basin, the main contributor to the Ribeirao do Onça, an important tributary of the Rio das Velhas basin, there are 64 streams and 280 springs, 66 of which are drained or landfilled. The area of the basin corresponds to around 20% of Belo Horizonte and is largely located in the sub-basin of the Isidoro stream. One of these streams, the Córrego dos Macacos, is considered the last clean watercourse in Belo Horizonte, is in an area of occupation, the focus of real estate speculation and is part of the Isidoro stream basin, which flows into the Onça stream, which makes up the Rio das Velhas basin, the target of conservation programs in the state. The vegetation and the number of local watercourses create a microclimate that is different from other areas of the city.

3.4 Weather

The state of Minas Gerais has a tropical transitional climate, influenced by large-scale global circulations and transient systems (dynamic factors) that interact with the regional topography, as well as with the diurnal cycle of temperature and humidity. Although a large part of Minas Gerais has this characteristic, it basically has two distinct seasons - one dry and one rainy. The state also varies greatly in terms of the spatial distribution of precipitation, due to its large size and the very steep terrain in some parts (CUPOLILLO, 1997).

The capital of Minas Gerais has a high-altitude tropical climate with a well-characterized rainfall regime with a rainy season from October to March and a dry season from April to September. According to figure 7, the graph representing the climatology of two 30-year periods of data analyzed respectively from the National Meteorological Institute (INMET/5°DISME), shows that the November to January quarter has the highest accumulation of rainfall, thus configuring two well-defined seasons.

Figure 7. Climatology of Belo Horizonte (MG)

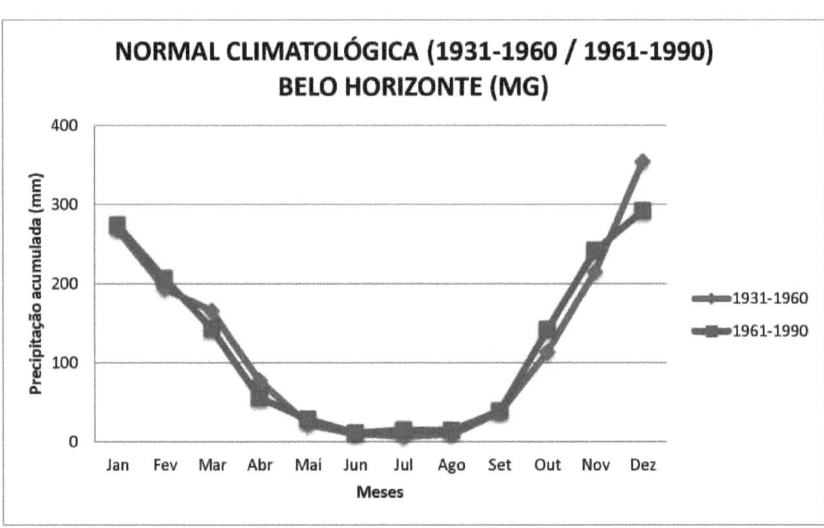

Prepared by the authors, based on data from INMET (National Institute of Meteorology). Station: 835887.

During the rainy season, more heating prevails over the continent, resulting in the formation of the Bolivian High. It also establishes a humidity corridor from the Amazon region to central and southeastern Brazil. In the November to January quarter, the South Atlantic Convergence Zone (SACZ) is common, a system resulting from the association of the aforementioned humidity corridor with the stationary positioning of a frontal system over the ocean, close to the coast of the Southeast. The SACZ consists of a band of cloudiness oriented in a northwest-southeast direction that keeps the area under which it forms with continuous rainfall for at least 4 consecutive days, being the main system responsible for the maximum rainfall observed in the November to January quarter. During the dry season, the South Atlantic Subtropical High approaches the continent, thus predominating a high-pressure system that inhibits rainfall in the Belo Horizonte metropolitan region (PRADO *et al.,* 2017).

The volume of annual rainfall in Belo Horizonte is approximately 1496 mm, with an average frequency of more than 100 rainy days (PRADO; GEMIACKI, 2016). The variability of rainfall has a social and economic impact on people's lives, and can cause disruption such as landslides, flooding, prolonged drought and low levels of relative humidity.

4. IMPACTS AND PERCEPTION OF THE ENVIRONMENT

3.5 . Impacts on the landscape

From an analysis of the images below, dated between 2011 and 2012 (1 year and 2 months), we can see the beginnings of landscape change in Granja Werneck, more precisely to the northwest, due to the anthropogenic action that gave rise to the Rosa Leâo community. These transformations intensified with the emergence of the two new communities - Esperança (the study area) and Vitória.

Figure 8. Northwest portion (2011)

Source: Google Earth

Figure 9. Northwest portion (2012)

Source: Google Earth

With its abrupt relief, part of the area is a water recharge zone with several springs, soil naturally prone to mass movement, cerrado and Atlantic forest vegetation, as well as a diverse fauna.

In view of the approaches taken and considering the reality of the area under study, it can be seen that among the existing socio-environmental problems, the most obvious are those directly linked to water issues (surface and underground), erosion, inappropriate cutting of slopes, solid waste (organic and inorganic) and the suppression of vegetation.

As the circumference of the area is one of intense residential activity, the direct or indirect interference caused by the use of this site is notorious, especially when it comes to the evolution of the gully that delimits the two communities in the region - Esperança and Vitória. The analysis of satellite images shows that before the insertion of the community, gullies had already started to form, and this quickly evolved into a large gully, causing a loss of material which probably silted up the watercourse downstream.

Figure 10. View of the gully in the region

Source: Google Earth (Jan. 30, 2012)

Figure 11. Gully (photo A)

Source: The authors (Nov. 2017)

Figure 12. Gully (photo B)

Source: The authors (Nov. 2017)

Figure 13. Gully (photo C)

Source: The authors (Nov. 2017)

Figure 14. Gully (photo D)

Source: The authors (Nov./2017)

In order to understand vulnerability, as cited by Esteves (2011) earlier in this paper, we can see densification near the banks of the gully, which can pose a risk to human life. The suppression of vegetation increases the potential for erosion and the improper disposal of solid waste of various kinds can lead to environmental problems.

The landscape shown in Figures 15 and 16 shows a suppression of vegetation *in situ,* which triggers a greater transport of sediment that is deposited downstream in a dynamic balance of the relief.

Figure 15. Silted up house (photo A)

Source: The authors (Nov. 2017)

Figure 16. Silted up house (photo B)

Source: The authors (Nov. 2017)

Regarding the Rua Bela landscape (figures 17 and 18), it requires priority care as it is on a steep slope and the source material is more susceptible to erosion.

Figure 17. Rua Bela (photo A)

Source: The authors (Nov. 2017)

Figure 18. Rua Bela (photo B)

Source: The authors (Nov. 2017)

Figure 19. Surface runoff

Source: The authors (Nov./2017)

Figure 20. Deposited material

Source: The authors (Nov./2017)

Various projects have been carried out in the region, all of which are palliative in nature. The PUC Minas Architecture university extension project uses social technologies with a sustainable bias, looking for resources that exist in the community itself and with low-cost materials, to mitigate erosion and the disposal of domestic effluent.

When it comes to water analysis, the focus of this study, CONAMA resolution 357/2005 points out reference values for water samples as portability after simplified treatment and the possibility of using the water for irrigation of vegetable crops, as expressed by the residents of Ocupaçao Esperança during the first field visit, requires particularly good quality. Nonetheless, several representative parameters of the samples indicate a quality level lower than that compatible with the intended uses (table 1).

Table 3: Values for the first water collection (May/2017)

WATER ANALYSIS IN CÓRREGO DOS MACACOS (COLLECTION 1)

PARAMETERS		**SAMPLES**			
Name / Unit	**Ref. values (CONAMA)**	**A**	**B**	**C**	**D**
pH	6,0 - 9,0	7,3	7,1	7,3	7,3
Temperature (°C)		25,8	25,9	25,9	26,1
BOD (mg O2t)^{1}	< 5	43,18	28,55	26,91	17,40
Suspended solids (mg 1)$^{-1}$		61,05	40,78	12,55	31,11
Total solids (mg 1)$^{-}$		1492,72	935,31	1113,85	868,02
OD (mg O2t)^{1}	> 5	0,47	1,72	1,30	3,95
Conductivity (µS⁄cm)		6,81	7,62	7,01	7,57
Turbidity (NTU)	< 100	282,37	223,49	309,49	198,09
Ammoniacal N (mg 1^)^{1}		4,17	3,79	3,73	3,06
P (MG 1^{-} 1)	< 0,025	0,318	0,192	0,266	0,209

Created by the authors, based on data collected.

In particular, the high BOD values and low DO values indicate the presence of excessive fractions of organic matter in the medium. These preliminary results, combined with values for nitrogen compounds and phosphates, significantly reduce the indication of direct consumption or followed by simplified treatment, leaving the categorization of the samples close to class 3.

Table 4: Values for the second water collection (November/2017)

WATER ANALYSIS IN THE CÓRREGO DOS MACACOS (COLLECTION 2)

PARAMETERS		**SAMPLES**			
Name / Unit	**Val. Ref. (CONAMA)**	**A**	**B**	**C**	**D**
pH	6,0 - 9,0	7,2	7,1	7,3	7,1
Temperature (°C)		28,2	28,8	29,5	28,6
BOD (mg O2t)^{1}	< 5	47,93	35,01	38,32	27,88
Suspended solids (mg 1)$^{-1}$		81,63	52,47	49,12	62,14
Total solids (mg 1^)^{1}		1981,43	1508,22	1464,93	1339,64
OD (mg O2t)^{1}	> 5	2,07	2,90	3,18	7,68

Conductivity (µS/cm)		5,28	6,79	5,44	5,17
Turbidity (NTU)	< 100	267,04	180,93	289,22	142,46
Ammoniacal N (mg 1^)^{1}		4,92	4,88	4,01	3,68
P (MG 1^{-} 1)	< 0,025	0,337	0,231	0,318	0,250

Created by the authors, based on data collected.

The sampling interval, although short, indicates positive and negative increases in some of the parameters that make up the IQA. It is still necessary to determine, in a broader sample, what role the variation in rainfall rates plays, the intensification of operations that characterize occupation in the community and the contribution of the communities surrounding the Esperança Occupation. It will also determine the dynamics of these parameters, especially considering the hydrodynamics of groundwater in the region.

The data show that especially the parameters N ammonium and P, both of which varied positively in the set of samples, point to the contamination of soils and groundwater by effluents from domestic waste.

3.6 Perception of the environment

During the interview, the residents' satisfaction with a palliative solution found by PUC Minas architecture undergraduates was evident. This solution consists of a small pit, filled with branches, and planting banana trees around it, called "ciclo de bananeira", to serve as a sink and toilet drain.

There was a unanimous response that the water used in the homes comes from the Copasa network, collected in a non-formal way, which reinforces the discourse of some of the interviewees on the practices of the residents in preserving the springs in the community area.

With regard to taking better care of the environment, it is clear that opinions differ, as some agree that there are environmental practices and others report that even with the mobilization that takes place at the meetings, some residents do not take up the cause of preserving the surroundings where most of them have lived since its inception in June 2013, To this end, the leaders are mobilizing to organize work parties to collect and dispose of the waste produced in the houses, which is taken to dumpsters installed in streets outside the community, which are served by Santa Luzia's urban collection service.

Due to the length of time they had been occupied, the residents hoped for more improvements, although they agree that many benefits will come with time, such as crèches, health centers, piped water, electricity, schools, regularization of land ownership, addresses, urbanization, churches, in all the speeches it is perceived that the residents' concern is to have a reference as citizens, as an address to have their basic needs met, as reported by interviewee A.

"[...] in order to be able to buy and have the product delivered, we need to give the address of the people who live in the Baronesa neighborhood, SAMU doesn't even come here when we call. This makes a lot of people here ashamed to say where they live [...]" (Interviewee A)

For the most part, the interviewees are people who came from other neighborhoods in the capital to get out of renting, and there is also a portion from Santa Luzia. With regard to the number of residents per home, it can be seen that they have an average of 3 to 4 residents, characterizing an urban reality.

FINAL CONSIDERATIONS

In carrying out this work, we realized that the object under study is much more complex than we had imagined. Land regularization and the urbanization of deprived communities are gaining prominence, given the need to establish more dignified housing and development conditions for the people who live there.

In view of the situation in the region, a change of zoning is suggested, which will allow the conflict to be resolved. It is also suggested that a Specific Global Plan (PGE) be drawn up for the Granja Werneck area, the main aim of which is to identify ways of improving the quality of life in these areas and integrating them into the city as a whole. One of the advantages of the Specific Global Plan is that it allows for greater efficiency and organization in the application of investments to avoid waste. Another is to raise funds from financial organizations (PBH, 2010).

The results of the water samples collected at some points upstream and downstream in the Córrego dos Macacos showed the presence of solid waste from improper disposal. Since the whole community wants to use the water abstracted from the Córrego dos Macacos, it is suggested that it be used for certain types of crops that are less demanding of drinking water quality (root crops - carrots, cassava, etc.). It is therefore proposed that there should be immediate mobilization focused on the problem of inadequate waste disposal with a methodology that covers it more effectively. It is important to emphasize the importance of the State intervening in the process of evolution of this natural phenomenon, providing security for residents.

According to Von Sperling (2005) "water quality is the result of natural processes and human intervention". Thus, in this study, we believe that most of the environmental stress in Granja Werneck is due to the lack of basic sanitation, which is not adequately provided by the municipality, the irregular disposal of solid waste and domestic effluents, chemical agents and deforestation are factors that influence surface runoff and the natural water cycle, causing alterations in chemical and biological structures. The soil, flora, fauna and surrounding human communities are closely linked to and dependent on this environment. It can also be deduced that the analysis of water in a given area requires temporal monitoring in order to better understand its dynamics, given that the climate shows seasonal and inter-annual variability and interacts with the soil, with consequences for the water table.

Of course, this work has some limitations, but we believe that it is possible to delve deeper into the issues raised here, as well as to use it for the community's dignified and legitimate struggle for access to housing with better conditions for human development, thus achieving a balance between society and nature.

ANNEX A

ANALYSIS PARAMETERS FOR WATER QUALITY

Parameters	Features
Color	Responsible for the color of the water.
Turbidity	Represents the degree to which light interferes with the landscape through the water, giving it a cloudy appearance.
Taste and smell	Taste is the interaction between taste (salty, sweet, sour and bitter) and smell (olfactory sensation).
Temperature	Measuring heat intensity
pH	Hydrogen potential. Represents the concentration of hydrogen ions H+, giving an indication of the acidity, neutrality or alkalinity of the water. The pH range is 0 to 14.
Alkalinity	Amount of ions in water that react to neutralize hydrogen ions. It is a measure of water's ability to neutralize acids (ability to resist pH changes: buffer capacity). The main constituents of alkalinity are bicarbonates (HCO3-), carbonates (CO3-) and hydroxides (OH-). The distribution of these three species in water is a function of pH.
Acidity	Water's ability to resist pH changes caused by bases. It is mainly due to the presence of free carbon dioxide (pH between 4.5 and 8.2).
Hardness	Concentration of precipitated bivalent cations.
Iron and manganese Chlorides Nitrogen Phosphorus	Suspended or dissolved solids
Dissolved oxygen	Dissolved oxygen (DO) is of essential importance for aerobic organisms (which live in the presence of oxygen). During the stabilization of organic matter, bacteria use oxygen in their respiratory processes, which can cause a reduction in its concentration in the environment.
Organic matter	Organic matter, present in bodies of water and sewage, is a characteristic of primary importance, causing the main problem of water pollution, the increase in dissolved oxygen (DO), caused by microorganisms in their metabolic processes of utilization and oxidation of organic matter.
Inorganic micropollutants	A large proportion of inorganic micropollutants are toxic. In especially heavy metals. Heavy metals that dissolve in water include arsenic, cadmium, copper, lead, mercury and silver. Several of these metals are concentrated in the food chain, resulting in great danger for organisms at the higher levels.

Organic micropollutants	Some organic materials are resistant to biotic degradation, becoming part of biogeochemical cycles and accumulating at a certain point in the cycle (interrupted). These include pesticides, some types of detergent (ABS, with a closed molecular structure) and a large number of chemical products. A large proportion of these compounds, even in low concentrations , are associated with toxicity problems.
Biological parameters	Indicator organisms / Algae / Bacteria

Adapted from VON SPERLING (2005)

REFERENCES

ARAÙJO, M. G. **The remaining quilombo community of Engenho Siqueira: territoriality, quilombola identity and the potential of agroecology.**

CUADERNOS DE GEOGRAFÌA | COLOMBIAN JOURNAL OF GEOGRAPHY. 2012, p. 99-114.

BRAZIL. **National Environmental Council**. Resolution No. 001, of January 23, 1986. Provides for basic criteria and general guidelines for the Environmental Impact Report (RIMA). Diàrio Oficial [da] Repùblica Federativa do Brasil, Brasilia, DF, 17 feb. 1986.

BRAZIL. **CONAMA Resolution on the Classification and Environmental Guidelines for the Classification of Groundwater**. Process No. 02000.003671/2005 71.

Available at: <http://www.mma.gov.br/port/conama/processo.cfm?processo=02000.003671/ 005-71>. Accessed on 13.05.2017.

BRAZIL. **National Environmental Policy**. Law 6.938, of August 31, 1981.

Available at: <http://www.planalto.gov.br>. Accessed on May 4, 2017.

BRAZIL. **National Water Resources Policy**. Federal Law No. 9.433, January 8, 1997. Available at <http://www.planalto.gov.br/ccivil_03/LEIS/L9433.

htm>. Accessed on April 9, 2017.

CETESB (Sao Paulo). **Report on Groundwater Quality in the State of São Paulo for the period 2004 - 2006 / CETESB**. Sao Paulo: CETESB, 2007. 199 p.

(Reports Series). Available at <http://www.cetesb.sp.gov.br/Solo/relatorios.asp>. Accessed on 13.09.2016.

CREPANI, et. al. **Remote Sensing and Geoprocessing Applied to Ecological-Economic Zoning and Territorial Planning**. Sao José dos Campos: INPE. 2001

CORRÊA, Roberto Lobato. **The urban space**. Sao Paulo: Atica publishing house, 1993.

COELHO, Maria Célia Nunes. **Environmental Impacts in Urban Areas**. In: GUERRA. Antônio José Teixeira; DA CUNHA, Sandra Baptista. Impactos Ambientais Urbanos no Brasil, 6th ed. Rio de Janeiro: Bertrand Brasil, 2010.

CUPOLILLO, **F. Periods of drought during the rainy season in the State of Minas Gerais:** spatialization and related dynamic aspects. 1997. Dissertation (Master's Degree in Agricultural Engineering) - Federal University of Viçosa, Viçosa, 1997.

DIAS, Maria Tereza Fonseca et al. Urban occupations in Belo Horizonte and the spatial (re)construction of the city: a case study of the Camilo Torres occupation. **Revista de Ciências Humanas**, Florianópolis, v. 49, n. 2, p. 205-223, nov. 2015. ISSN 2178-4582. Available at: <https://periodicos.ufsc.br/index.php/revistacfh/article/view/2178- 4582.2015v49n2p205>. Accessed on: June 4, 2017.

DOLLFUS, Olivier. **O espaço geogràfico** (translated by Heloysa de Lima Dantas) Sao Paulo - Rio de Janeiro, 1975. Editora Difel. 2 ed. p. 1-121. Current Knowledge Collection.

EGLER, C. A. G. **Environmental Risk as a Criterion for Territorial Management**. Territòrio, 1, pp. 31-41. 1996.

ESTEVES, Clàudio Jesus de Oliveira. **Risk and Socio-environmental Vulnerability**: Conceptual Aspects. Cad. IPARDES. Curitiba, PR, v.1, n.2, p. 62-79, jul./dez. 2011.

FAUSTINO, J. **Planificación y gestión de manejo de cuencas**. Turrialba: CATIE, 1996. 90p.

FOUCHER, M. **Esquisse d'une Géographie Humaine des Risques Naturels**. Hérodote, 24: 40-67. 1982.

FRANCISCO, Wagner de Cerqueria e. Reforma **Agrària**; Brasil Escola. Available at <http://brasilescola.uol.com.br/sociologia/reforma-agraria.htm>. Accessed on November 22, 2017.

GARCIA, Ricardo Alexandrino. RIBEIRO, Adriana Miranda. Socio-spatial segregation in Belo Horizonte: **an application of fuzzy models**. Available at <http://www.abep.nepo.unicamp.br/iussp2001/cd/GT_Pop_Trab_Ribeiro_Garcia_Text.pdf>. Accessed on Aug. 30, 2016.

GARCIAS, Carlos Mello. SANCHES, Alexandre Martinho. Socio-environmental vulnerabilities and urban water availability: **conceptual theoretical survey and analysis applied to the Metropolitan Region of Curitiba - PR**. Revista de Pesquisa em Arquitetura e Urbanismo - EESC USP. 2009.

GONÇALVES, Thiago GilibertiBersot. Segregated peripheries, segregation in the peripheries: **an analysis of intra-urban inequalities in the municipality of Sao Gonçalo, RJ**. Available at <http://www.observatoriodasmetropoles.net/download/periferia_bersot.PDF>. Accessed on September 16, 2016.

GOTTSCHALG, Maria de Fâtima S. Urban Socio-Spatial Segregation and State Intervention: **A geographic-social approach**. Available at <http://www.cress-mg.org.br/arquivos/DE%203.pdf>. Accessed on September 30, 2016.

HORTON, R. E. **Erosional development of streams and their drainage basins**: hydrographical approach to quantitative morphology. Geol. Soc. Am. Bull. v. 56, n. 3, p. 275-370, 1945.

JANSEN, D.C. **Environmental analysis of the Morro da Pedreira Environmental Protection Area and the Serra do Cipó National Park for the Protection of Speleological Heritage**. 2θl3. 149f. Dissertation (Master's Degree in Geography - Spatial Information Treatment) - PUC Minas. Belo Horizonte. Available at: <http://www.biblioteca.pucminas.br/teses/TratInfEspacial_JansenDC_1.pdf>. Accessed on: Aug. 13, 2017.

LEFEBVRE, Henri. In the light of dialectical materialism. Logique formelle, logique dialectique. F. Paris, Editions Sociales, 1947 *apud* GEIGER, Pedro P. **Território;**

globalization and fragmentation. Sao Paulo, 1998. Editora Hucitec.

LIMA, Adalto Gonçalves. The hydrographic basin as a field of study in Human Geography. **Geography Magazine - UEL**. V. 14, N° 2, Jul - Dec. State University of Londrina - UEL. 2005.

MALDI, D. (1997). **From confederates to barbarians: the representation of indigenous territoriality and frontiers in the 18th and 19th centuries.** *Revista de Antropologia* , 183-221.

MACHADO DA SILVA, Luis Antônio. **"The continuity of the 'favela problem"**. In: OLIVEIRA, Lùcia Lippi (org.) **Cidade: história e desafios**. Rio de Janeiro: Editora FGV, 2002.

MENDONÇA, F. A. Riscos, Vulnerabilidade e abordagem socioambiental urbana: uma reflexao a partir da RMC e de Curitiba. **Desenvolvimento e Meio Ambiente**, Curitiba: Ed. da UFPR, n.10, p.139-148, jul./dez. 2004.

MOREIRA, Antônio Clàudio M.L. **Conceitos de ambiente e impacto ambiental aplicàveis ao meio urbano.** Sao Paulo: 1997.

MOREIRA, Antônio Clàudio M. Lima. Mega-projects & the Urban Environment: a methodology for preparing a Neighborhood Impact Report. **Revista do Programa de Pós Graduaçâo da FAU/USP**, Sao Paulo, n. 7, p. 107-118, 1999. Urban planner.

PRADO, G.M.S.; L, GEMIACKI. **Quantitative Analysis of the Number of Rainy Days in Belo Horizonte in the Historical Series:** 1910 to 2015. In: BRAZILIAN CONGRESS OF METEOROLOGY, XIX, 2016. JoaoPessoa, PB.Anais... Joao Pessoa: SBMET 2016.

PRADO, G.M.S.; L, GEMIACKI; FERNANDES, A.S. **Occurrences of Veranicos in Belo Horizonte from 2007 to 2017**. In: SIMPÒSIO INTERNACIONAL DE CLIMATOLOGIA, VII., 2017. Petrópolis, RJ. Proceedings... Petrópolis: SBMET 2017.

SANTOS, Milton. **Metamorphoses of inhabited space. Theoretical and methodological foundations of Geography.** Sao Paulo, 1997. Editora Hucitec, 5ª Ed. P. 1-124.

SANTOS, Milton. The nature of space: **technique and time, reason and emotion**. 4. ed. Sao Paulo: EDUSP, 2008.

SAQUET, Marcos Aurélio. **Approaches and Conceptions of Territory.** Sao Paulo: Expressao Popular, 2007.

SILVA, Luciene Pimentel. Hydrology: **engineering and the environment.** Sao Paulo: Elsevier/Campus, 2015

SCHIAVETTI, A., CAMARGO, A. F. M. (Ed.). Watershed concepts: **theories and applications.** Ilhéus, BA: Editus, 2002.

SUZUKI, Jùlio César. **The agrarian question in Latin America:** capitalized income as a tool for reading socio-spatial dynamics. In publication: Latin America: city, countryside and tourism. Amalia Inés Geraiges de Lemos, Mónica Arroyo, Maria Laura Silveira. CLACSO, Consejo Latinoamericano de Ciencias Sociales, San Pablo. December 2006.

VON SPERLING, M. **Introduçao à qualidade das águas e ao tratamento de esgotos.** Belo Horizonte: Department of Sanitary and Environmental Engineering. Belo Horizonte: Ed. da UFMG, 2005.

TAKEDA, Tatiana. **Land Use and Occupation.** Available at <https://www.jurisway.org.br/v2/dhall.asp?id_dh=12363> Accessed on: May 23, 2017.

TRICART, Jean. **Ecodynamics**. Rio de Janeiro. IBGE, Technical Directorate, SUPREN, 1977.

TEODORO, V. L. I.; TEIXEIRA, D.; COSTA, D. J. L.; Fuller, B. B. **The watershed concept and the importance of morphometric characterization for understanding local environmental dynamics**. Revista Uniara, v.20, p.137-157, 2007.

TUCCI, Carlos E. M. Aguas urbanas. **Revista Estudos Avançados**, Sao Paulo, v. 22, n° 63, 2008.

TUNDISI, José Galizia et al. **The Tietê/Jacaré river basin: a case study in research and management**. Estudos Avançados, Sao Paulo, v. 22, n. 63, p. 159172, jan. 2008. ISSN 1806-9592. Available at: <http://www.revistas.usp.br/eav/article/view/10298/11949>. Accessed on: June 4, 2017.

VON SPERLING. **Introduction to water quality and sewage treatment**. Volume 1. DESA. UFMG. 2006.

Printed by Books on Demand GmbH, Norderstedt / Germany